Lioudmila Aleksandrovna Matlakhova
Oleg Konstantinovich Belousov
Natalia Alekseevna Palii

Elastic Modulus and Structure Changes in TiNi during SME and RSME

Lioudmila Aleksandrovna Matlakhova
Oleg Konstantinovich Belousov
Natalia Alekseevna Palii

Elastic Modulus and Structure Changes in TiNi during SME and RSME

LAP LAMBERT Academic Publishing

Imprint

Any brand names and product names mentioned in this book are subject to trademark, brand or patent protection and are trademarks or registered trademarks of their respective holders. The use of brand names, product names, common names, trade names, product descriptions etc. even without a particular marking in this work is in no way to be construed to mean that such names may be regarded as unrestricted in respect of trademark and brand protection legislation and could thus be used by anyone.

Cover image: www.ingimage.com

Publisher:
LAP LAMBERT Academic Publishing
is a trademark of
International Book Market Service Ltd., member of OmniScriptum Publishing Group
17 Meldrum Street, Beau Bassin 71504, Mauritius

Printed at: see last page
ISBN: 978-613-9-85937-5

Elastic Modulus and Structure Changes in TiNi during SME and RSME

L. A. Matlakhova, O. K. Belousov, N. A. Palii

Dedication

The Authors dedicate this work, expressing their deep appreciation and gratitude for the great contribution to its implementation, to:

Yuliy (Juli) Konstantinowitsch Kovneristy *(1927-2007), Doctor of Technical Sciences, Academician of the Russian Academy of Sciences (RAS), Director of the IMET RAS,*

Serguei Gerasimovich Fedotov *(1921-2004), Leading Researcher at IMET RAS, Ph.D.,*

and

Anatoliy Nikolaevich Matlakhov *(1945-2011), Leading Researcher at MISIS, Ph.D. Associate professor of UENF, Campos, RJ, Brazil*

Preface

The direct (SME) and the reversible (RSME) shape memory effects are governed by structural changes caused by previous plastic deformation of alloys that undergo reversible martensitic transformations (RMT). This paper place high emphasis on the effect of prior plastic deformation by tension ($\varepsilon_0 = 2 \div 30\%$) on the structure, electrical resistivity, dynamic elastic modulus E along with SME/RSME and critical temperatures of RMT, using TiNi alloy with initial martensitic structure B19$'$. Key features of SME/RSME and the critical temperatures of RMT were determined by dilatometry and by DTA respectively. The values of E were determined at room temperature (RT) and during heating-cooling cycles using Elastomat equipment. The structural changes were analyzed by XRD. The tension tests were performed in an Instron machine at RT. The results show that in the RMT interval, during heating or cooling, E goes through minimum (E_{min}), indicating the loss of elastic stiffness of the original phase. The effect of ε_0 on the structure, E, SME, and RSME was determined. At the pseudo-yield plateau the alloy suffers B19$'$ to R/B2 phase transformation, and E/E_{min} show anomalous behavior with previous residual deformation. Systematic studies allowed us to develop/create the original ideas on the processes that accompany the accumulation and recovery of reversible deformation in SME/RSME realization.

The authors are grateful to **Svetlana Vasilievna Oleynikova**, and **Natalia Fiodorovna Zhebynieva** for useful discussions and for the invaluable technical support of the work.

Author Contributions

L.A. Matlakhova wrote the manuscript, performed tension tests, moduli tests and dilatometry analysis (with participation of S.G. Fedotov and orientation by Y. K.

Kovneristyi), the XRD analysis (with technical support of S.V Oleynikova and A.N. Matlakhov), resistivity tests (with participation of N.A. Palii).

O.K. Belousov performed DTA tests, moduli RT tests, and participated in discussion of results.

N.A. Palii performed resistivity tests, participated in discussion of results, as well as translation from Russian into English and article editing.

List of notations

SME	Shape Memory Effect
RSME	Reversible Shape Memory Effect
SE	Superelasticity
SMA	Shape Memory Alloy
MT	Martensitic Transformation
RMT	Reversible Martensitic Transformation
RT	Room Temperature
DTA	Differential Thermal Analysis
XRD	X-ray Diffraction
ε_o	Prior Plastic Deformation
E	Dynamic Elastic Modulus
G	Shear Modulus
ρ	Electrical Resistivity
A	Austenite
M	Martensite
M_{ind}	Induced martensite
A_f	Austenite Finish Temperature
A_s	Austenite Start Temperature
M_f	Martensite Finish Temperature
M_s	Martensite Start Temperature
R_s	R-phase Start Temperature at cooling
R_f	R-phase Finish Temperature at cooling
M_d	Martensite Stress-Induced Critical Temperature
A_d	Austenite Stress-Induced Critical Temperature

T_d — Deformation Temperature

E_{min} — Elastic modulus minimum

T_{Emin} — Temperature of E_{min}

E_{int} — Elastic modulus at the end of the intensive rise when heated

T_{Eint} — Temperature of E_{int}

$A/\sigma \rightarrow M_{ind}$ — Austenite to induced martensite transition under strain

$M/\sigma \rightarrow \beta'_{induced}$ — Martensite to $\beta'_{induced}$ transition under strain

$\beta'_{induced} \rightarrow M_{reoriented}$ — $\beta'_{induced}$ to reoriented martensite transformation

Content

1. Introduction

To understand the nature and peculiar features of alloys with inelastic effects, such as Shape Memory Effect (SME), Reversible Shape Memory Effect (RSME), Superelasticity (SE), it is important to study structural changes along with structure-sensitive properties while implementing these effects, i.e. under loading (deformation), unloading, heating and cooling.

First observed in the Au-47.5 % Cd alloy by Chang and Read, in 1951 [1], the unique behaviour of SME was publicized with the revelation in TiNi alloys by Buehler et al., in 1963 [2]. For more than half a century since discovering SME, significant achievements in understanding of these effects have been made. It was established that SME, RSME, and SE are associated with the special "thermoelastic" (or reversible) type of martensitic transformations (MT) discovered by Kurdyumov and Khandros [3] in 1949, when studying intermetallics in Cu-Al-Ni and Cu-Zn systems. The similar nature of MT was observed in Cu-Cd, In-Tl, Cu-Zn, Cu-Al-Ni, Fe-Mn-Si, Ti-Nb, and Ti-Ta alloys; all of them display inelastic properties under appropriate processing [4-6].

Even though the alloy with RMT is deformed in the low temperature phase, it recovers its original shape by the reverse transformation upon heating over the critical temperature, called the reverse transformation temperature. So-called SMAs (shape memory alloys) likewise display SE and RSME. The effect of SE is associated with the large (up to 18 %) non-linear recoverable strain upon loading and unloading at the higher temperature. To enable RSME, it is necessary to create internal stresses that orient martensitic shifts at thermocycling of pre-deformed alloy [4 - 7].

SMAs, or RMT-alloys, exhibit: (a) coherent boundary between martensitic (M) and high-temperature (A) phases; (b) minor changes in Gibbs energy at reversible phase transformations A $\leftrightarrow$ M, about a few dozen of J/mol (for TiNi, $\Delta G^{o} \sim$ 67-83 J/mol, while in martensitic alloys without inelastic effects ΔG^{o} reaches 2400 J/mol (Fe-10 % C) and 1450 J/mol (Fe-10 % Cr); (c) zero or narrow temperature hysteresis of RMT,

$\Delta T < 50\ °C$; (d) inelastic effects, provided temperature stability of metastable martensitic and high-temperature phases (without decomposition and stable phases formation) in the quenched alloy; (e) similar behaviour of elastic moduli, i.e. abrupt decrease to the minimum in RMT critical temperature range, due to lattice elastic instability of the involved phases; (f) the significant changes in physical and mechanical properties, i.e. electrical resistivity, internal friction, sound velocity, thermal dilatation, magnetic susceptibility, etc. [**5, 9-15**]. Deformation curve of RMT-alloys has the characteristic shape (**Figure 1**) with distinct regions of: (0-a) - elastic deformation of the initial alloy structure, up to the first yield point; (a-b) – pseudo-yield plateau, where preferential pseudo-plastic deformation (restorable upon unloading (SE) or heating above A_f (SME)) is accumulated; (b-c) – mainly elastic deformation leading to the hardening of the structure formed at the pseudo-yield plateau; (c-d) - foremost plastic deformation up to destruction. At the pseudo-yield plateau the reversible deformation is accumulated via phase transformations and reorientations of the martensitic lamellae [**3-5, 13-20**]. Otsuka et al. [**21**] and Saburi et al. [**22**] have established the possible existence of multiple pseudo-yield plateaus, depending on the number of phase transformations in the alloy under loading, e.g. three $(\beta \rightarrow \beta'_1 \rightarrow \gamma'_1 \rightarrow \beta''_1)$ in Cu-14.0 mass % Al-4.2 mass % Ni [**21**], and two $(B2 \rightarrow R \rightarrow B19')$ in TiNi [**22**].

By deforming RMT-alloy within certain temperature intervals of direct $(M_s\text{-}M_f)$ or inverse $(A_s\text{-}A_f)$ martensitic transformations we may tune inelastic effects (**Figure 2**).

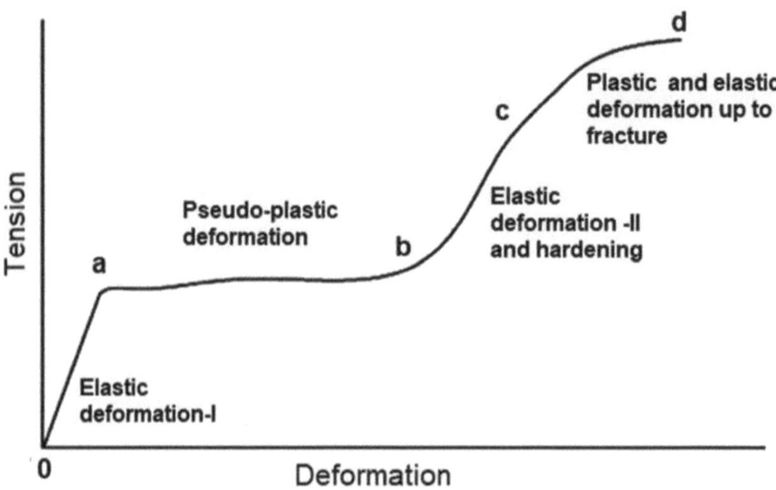

Figure 1. Typical deformation curve of TiNi SMA.

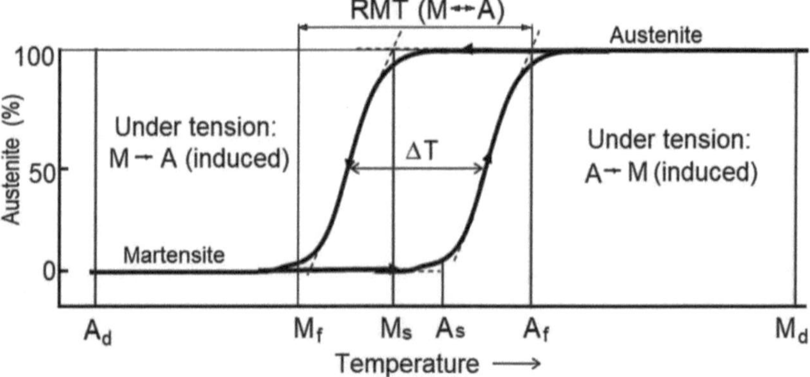

Figure 2. RMT scheme of SMA under tension; the critical temperatures on heating (A_s - A_f) and on cooling (M_s - M_f), the temperature hysteresis ΔT, and the critical temperatures M_d and A_d are indicated.

The austenitic (A) RMT-alloy, can exhibit superelasticity when deformed above A_f ($A_f <T_d<M_d$); thereby the loading induce the inelastic strain accumulation by stress-induced martensite oriented transformation ($A/\sigma \rightarrow M_{ind.}$), by this way a macro-deformation is accumulated in the alloy. $M_{ind.}$ is thermodynamically unstable in undeformed state in this temperature range, so M_{ind} transforms to the stable high-temperature austenite under unloading; in doing so macro-structure distortions are removed, the alloy takes its original shape. In (A_f-M_d) temperature interval, the higher the deformation temperature (T_d) the greater the strain must be applied (σ_{A-M}) to initiate the ($A/\sigma \rightarrow M_{ind}$) transformation [23]. Above the critical temperature M_d the induced martensite does not form, and, respectively, SE does not develop. So alloy deforms without phase transformations, its yield stress decreases with increasing temperature, exhibiting typical elastic-plastic behaviour of the original stable structure [23-25].

Within (A_s-A_f) temperature range, the applied stress also initiates the ($A/\sigma \rightarrow M_{ind.}$) transformation resulting in formation of stress-assisted martensite partially stabilized due to thermodynamics, thus preserving structure distortions; the accumulated macro-deformation is completely restored only when the deformed alloy is heated above A_f. In this interval, the effects of SME, RSME, and SE, can be observed simultaneously in different ratios.

In the M_f-A_f temperature interval, phase transformations under loading and deformation are quite complex, the involved phases are unstable; and the resulting strained structure depends on pre-history, i.e. whereby the alloy has reached the definite temperature: by heating from complete martensite state or by cooling from austenite (Figure 2) [26-28].

Different theories of deformation mechanism in SMAs, in a completely martensitic state, below M_f, were proposed, such as: reorientation of the martensite plates and micro-twinning [9, 13, 29], martensite-martensite transformations [30], stress-assisted double transformation [26, 31], martensite to the high temperature phase transition as a result of plastic deformation [32]. But there is still no consensus, the issue is under discussion. According to the double transformation mechanism by Wasilevski [26, 31], within the A_d–M_f temperature range, martensite can transform to the transition high temperature phase under stress (M/σ→$\beta'_{induced}$); while unloading, this instable phase may be–converted into martensite of a new, favorable orientation ($\beta'_{induced}$→$M_{reoriented}$). Owing to Wasilevski, the concept of the critical temperature A_d was introduced. Above this temperature, unstable in the (A_d-M_f) interval martensite can be transformed into stress-oriented high-temperature phase (or coherent transition phase) by RMT. At unloading such unstable phase becomes stable or turns into martensite that inherits structure orientation, preserving the macro-deformation. Under subsequent heating, reversible phase transformations remove these distortions and macro-deformation, so the alloy tends to recover the original shape.

The symmetric contraction of pseudo-yield plateau observed for the deformed within M_f - A_f critical interval RMT-alloys [24-26] (when the initial structure is whether martensite or high temperature phase or a mixture of both), supports Wasilewski's assumption. Although confirmed in a number of papers [17-19, 33], the formation of the high-temperature-induced-phase in RMT-alloys under deformation, so far raise debate.

The elastic properties of alloys are the fundamental characteristics of the binding energy of the lattice atoms [34] and are determined for small deviations of the atoms from their equilibrium positions. The modulus of elasticity (or Young's modulus), E, the alloy's stiffness, is a material's intrinsic property. The shear modulus, or modulus of rigidity (G) describes the tendency to shear (the deformation of shape at constant volume) when acted upon by opposing forces.

The decrease of elastic and shear moduli observed in the vicinity of the cubic to martensite phase transformation is typical for TiNi, **Figure 3** [10], as well as for Au-Cd [35], Ti-Mo, Ti-V, Ti-Nb [36, 37], and other SMAs, or RMT-alloys [38]; such the decrease occurs both on cooling to M_s and on heating to A_s, and the Poisson's ratio reaches the highest values in the critical interval [13, 35]. At the completion of the phase transitions the moduli grow [13, 15, 16, 35, 37-39].

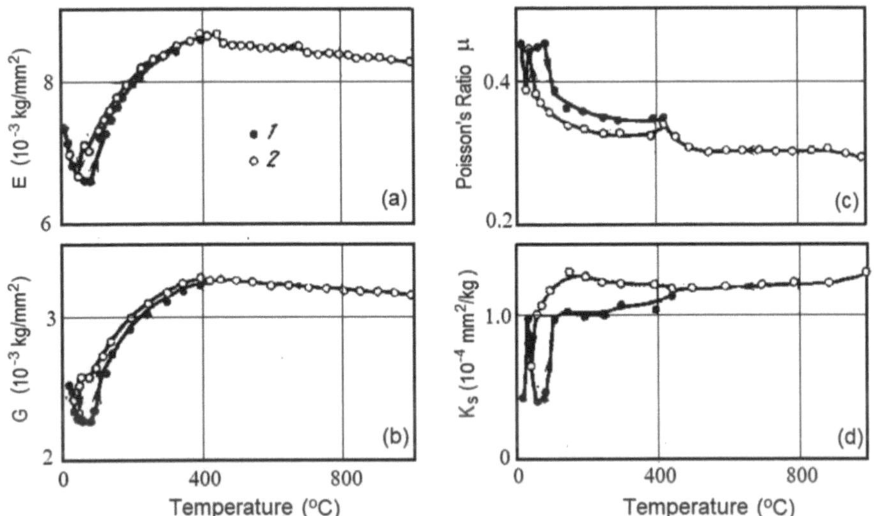

Figure 3. Temperature dependence of moduli E and G, Poisson's ratio μ and adiabatic compressibility K_s of TiNi alloy quenched from 980 °C: (1) • - on heating and (2) o - on cooling (2) [**10**].

In general, MT is driven by shear instability of the parent lattice [**39-43**], resulting in dramatic decrease of elastic moduli and increase in the Poisson's ratio upon approaching the critical points [**44**]. MT kinetics can vary, depending on the type of interaction of elements (determined by the state diagram) and on the different mobility of the atoms of the involved elements. In the case of intermetallics (TiNi, Cu_3Al, etc.) the inverse MT is almost hysteresis-less, provided martensite is stable up to A_f. These alloys tend to display SME and SE; MT propagates by reversible, or thermoelastic, mechanism there [**4-10, 14-15, 29**].

In this paper, we present more complete results, not published in the earlier papers [**16-19**], as well as the new data. Special attention is paid to the elastic properties of the alloy with martensitic structure in its initial and deformed state. **The aim of this work is to present** the effect of deformation by tension up to 30 % on the structure and properties (electrical resistivity, modulus E, key features of direct (SME) and reversible shape memory effect (RSME) and critical temperatures of RMT) of TiNi alloy with initial martensitic structure B19′.

2. Experimental section

2.1. TiNi alloy specimens preparation and analysis

Investigations were carried out using a rod of Ti- 54 wt. % Ni alloy ($\varnothing$20 mm), manufactured by All-Russia Institute of Light Metals and then treated at the Baikov Institute of Metallurgy and Materials Science, RAS, Moscow (IMET). Sections of the initial rod were heated to 900 ^{o}C and hot rolled into a bar with $\varnothing$12 mm. Then specimens were machined from the bar, and annealed in an evacuated (0,133 Pa) quartz ampoule for 1 hr at 500 ^{o}C then cooled in air (in ampoule). The samples of the TiNi alloy, with working part length 100 mm and $\varnothing$7 mm, were deformed, separately, by tension in an Instron-2351 testing machine at the rate 3.3×10^{-5} m/sec (with controlled unloaded) at RT, up to different levels of total tension (ε_{tot}), and unloaded up to the state without any elastic component, varying previous (extent) deformation (ε_{o}) from 2 % to 30 %, at RT . Cylindrical samples, for measuring the modulus of elongation, and specimens of disks, to perform DTA, XRD and dilatometric analysis, were electro-machined by electro-erosive cutting, from the working part of deformed specimens. Cutting scheme is presented in **Figure 4**.

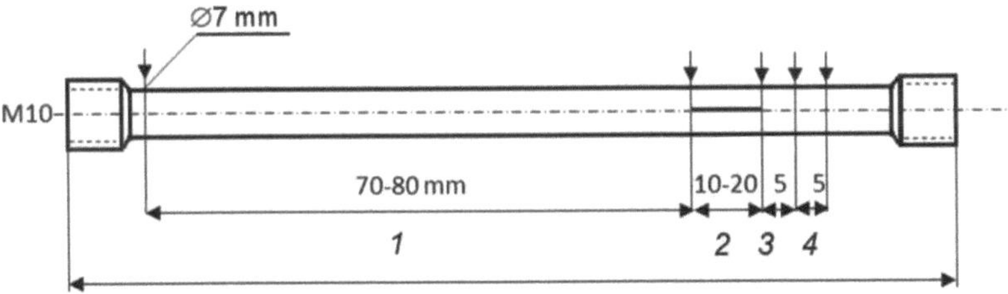

Figure 4. Scheme of cutting of the test specimens for: (1) elastic properties measuring, (2) thermal X-ray analysis, (3) dilatometry, and (4) RT X-ray analysis.

SME and RSME parameters were studied by dilatometry, during heating-cooling cycles. The RMT critical temperatures were determined by DTA. The dynamic elastic modulus was determined at RT and during heating/cooling cycles, using Elastomat equipment. The structure was analyzed by X-ray diffraction (at RT and during heating and cooling). In addition electrical resistivity and density were investigated.

2.2. Electrical resistivity measurement

Electrical resistivity is very sensitive to the structural changes of SMAs. The electrical resistivity (ρ) was determined using a Wheatstone bridge circuit with a constant current source (I); while measuring the voltage drop (V_x), electrical resistance ($R_x=V_x$ /I), cross-section area (S; $S = \pi d^2$, d - is diameter), and the length of the measured section (L). Finally, the resistivity was determined by the Equation (1):

$$\rho = R_x \times (S/L) \qquad (1)$$

2.3. Density measurement

The density of the alloy (γ_a), in its initial state and after deformation, was determined by hydrostatic weighing (Archimedes method) and using the Equation 2.

$$\gamma_a = \frac{P_{air} \cdot (\gamma_w - \gamma_{air})}{P_{air} - P_w} + \gamma_{air} \qquad (2)$$

The obtained data were used to calculate the E and G moduli, determined by dynamic resonance method.

2.4. Dilatometric analysis

The aim of the dilatometric analysis was to validate and investigate the effects SME and RSME as the function of the retain deformation (ε_o). It was performed at UIP-70M dilatometer (IMET), on heating and cooling, heating rate = 8.10^{-2} $^{\circ}$C/sec (5 $^{\circ}$C/min). Dilatometric curves were plotted in the "relative contraction, $(\Delta h/h_o)$ – temperature" coordinates, where h_o is the size of the deformed specimen, and Δh is the size change.

2.5. DTA

The differential thermal analysis (DTA) was realized to determine the MT critical temperatures of the investigated TiNi alloy, in its initial annealed and deformed states. DTA was performed in ATVU-10A machine (IMET) at the heating rate of 8×10^{-2} $^{\circ}$C/sec, using the TG00 titanium as a standard.

2.6. Measuring of modulus of elongation by dynamic method

The dynamic elastic modulus was determined at 20 $^{\circ}$C and during heating/cooling cycles using the Forster Elastomat equipment. The essence of the method is to measure the resonance frequencies of samples submitted to transverse, longitudinal, and torsional vibrations. The first two resonance modes, $f_{longitudinal}$ and $f_{transverse}$, allow to calculate the elastic modulus, or the Young's modulus, E; the third mode, $f_{torsional}$, makes possible the calculation of the shear modulus, G. The relations between these characteristic resonance frequencies follow the Equations (3, 4); the moduli E and G are calculated according to the Equations (5, 6):

$$f_{\text{longitudinal}} = 0.561 \cdot (L/d) \cdot f_{\text{transverse}} \qquad (3),$$

$$f_{\text{torsional}} \cong 0.6 \cdot f_{\text{longitudinal}} \qquad (4),$$

$$E = 4{,}0775 \cdot \gamma \cdot L^2 \cdot (f_{\text{longitudinal}})^2 \cdot 10^{-7} \ (MPa) \qquad (5),$$

$$G = 4{,}0775 \cdot \gamma \cdot L^2 \cdot (f_{\text{torsional}})^2 \cdot 10^{-7} \ (MPa) \qquad (6),$$

where γ - is the density (g/cm^3), L – is the length (mm), and d – is the diameter (mm) of the specimen.

The Elastomat forced vibration frequency range supplied to the specimen was 0.6 – 26 kHz, maximum tension was within 10^{-1} MPa, and the relative deformation was of order 10^{-3} - 10^{-5} %. Elastomat was equipped with a furnace for heating up to 1000 $^{\circ}$C. The moduli E and G of initially annealed and deformed specimens were determined at room temperature, afterwards both specimens were heated from RT to 300 $^{\circ}$C - 700 $^{\circ}$C at the rate of 8×10^{-2} $^{\circ}$C/sec, and cooled passively.

2.7. Structure analyses

Phase composition of the TiNi alloy in initial annealed and deformed state, was analyzed by a X-ray diffractometer DRON-2 (IMET, Moscow) in filtered Cu, K_{α}-radiation (λ = 0.154178 nm) in 2θ range of 36° to 48° at RT. The X-ray diffraction measurements also were carried out at temperatures from +250 $^{\circ}$C to -160 $^{\circ}$C, using a URS-50-IM diffractometer (VILS, Moscow), equipped with a high temperature attachment for heating to 400 $^{\circ}$C, and with the low-temperature attachment KRN-190. The temperature was controlled by the chromel-alumel thermocouple placed directly on the test sample. The testing space was evacuated during temperature investigations. The diffraction analyses were done in the angular range 2θ = 36-46° in filtered Cu, K_{α}-radiation with isothermal holding of about 10 min at the each temperature. Microstructure of the alloy was observed by optical microscope Neophot 32.

3. Results

3.1. Structure and properties of TiNi in initial annealed state

The structure of the TiNi alloy, in its initial annealed state, was B19′ martensite with fine inclusions of Ti_2Ni intermetallic according to optical microscopy. The moduli E and G were determined as 59-61 GPa and 21-23 GPa, respectively (at RT), as well as the electrical resistivity ρ, density and critical temperatures (DTA); the results are presented in the **Table 1**.

Table 1. Properties of the Ti-54 wt. % Ni alloy in its initial non deformed state

Structure	Modulus E, (GPa)	Modulus G, (GPa)	Electrical resistivity, ρ ($\Omega\times$cm)	Density, (g/cm^3)	A_s - A_f (^oC)	M_s − M_f (^oC)
B19′+ (Ti$_2$Ni)	59-61	21-23	(82-84)×10^{-6}	6.4	55-95	53-38

During heating and cooling, the initial annealed (non-deformed) TiNi alloy suffers B19′↔B2 martensitic transformation (**Figure 5a, b** and **Figure 6a**), B2↔B19′ RMTs are evidenced in the form of slight inflexions on the dilatation curves of undeformed TiNi alloy (**Figure 6b**); both phases suffer natural expansion (heating) and compression (cooling) in ranges of their existence.

B2-phase thermal expansion coefficient (α) was determined as 15.62×10^{-6} grad^{-1}, in the range of 200-300 ^{o}C.

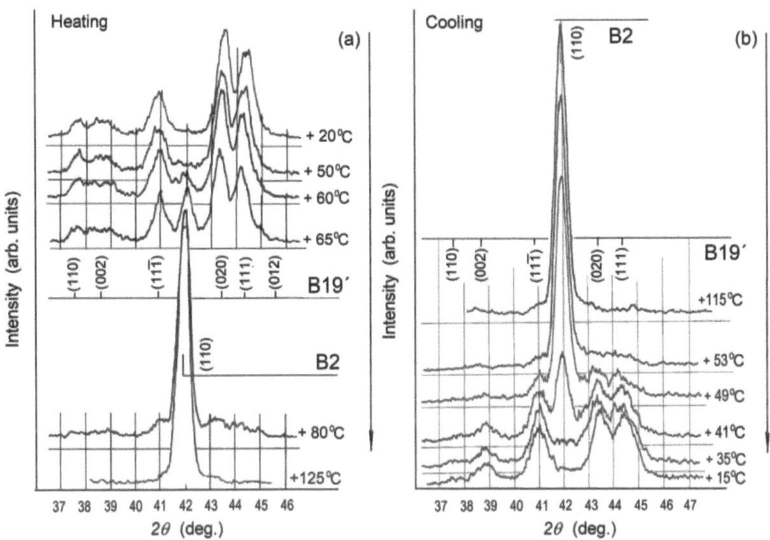

Figure 5. XRD of the non-deformed Ti-54 wt. %Ni alloy during: (a) heating and (b) cooling.

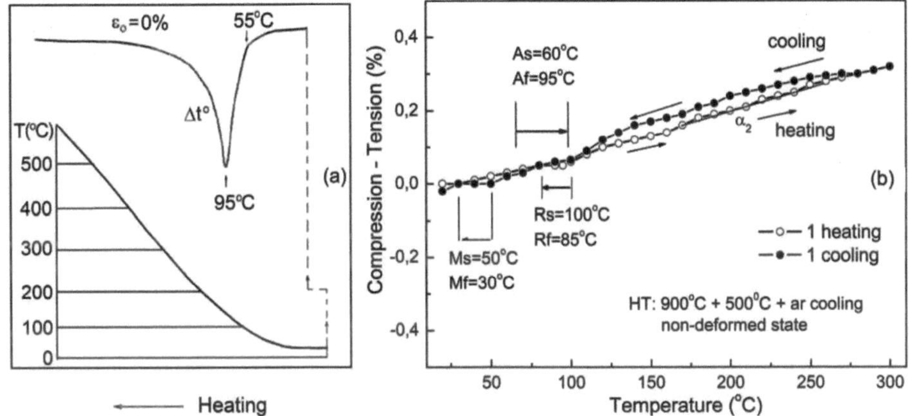

Figure 6. (a) DTA (heating) and (b) dilatometric analysis (heating and cooling) of Ti-54 wt. % Ni alloy, in the initial annealed, non-deformed state.

3.2. Moduli E and G of Ti-54 wt. %Ni on heating

On heating of the initial TiNi alloy, the modulus of elongation E and the shear modulus G are changing abnormally (**Figure 7**). Four temperature zones may be pointed out. The first zone is at T = 20 ÷ 60 °C, E is going down to 51-49 GPa i.e. to $E_{min.}$; the second one ranges from 60 °C to 100-110 °C, E is rising abruptly to 83-85 GPa, up to E_{int}; and the third one is at T > 110 °C, E is increasing gradually during heating up to 700 °C (96.4 GPa). The shear modulus displays similar anomalous behavior: initially G is going down to 20.3 GPa (60 °C), within the second zone G is rising up to ~25 GPa, and is increasing gradually achieving 31.5 GPa at 700 °C in the third zone.

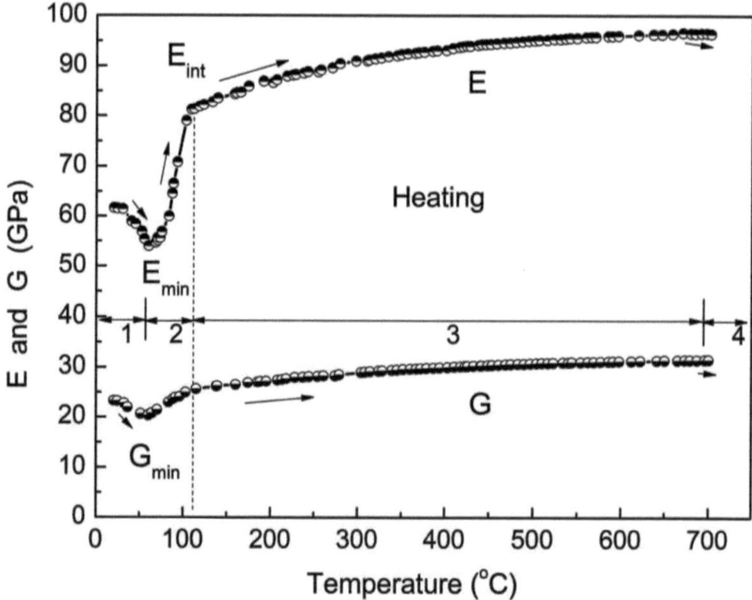

Figure 7. Moduli E and G of the non-deformed Ti-54 wt. %Ni alloy, during heating.

A Poisson's ratio (μ) reveals abnormal high values, up to 0.6, when the elastic and shear moduli decrease, so TiNi alloy manifests itself as anisotropic, inelastic material, especially in the RMT range (B19′-B2). But the notable decrease of Poisson's ratio during heating within the third zone may indicate B2-lattice stabilization at the higher temperatures. On cooling the moduli reversed, with temperature hysteresis equal to the same of RMT (**Figure 8 a**).

The reduction of the both moduli at 20-60 °C (**Figure 8a**), in the 1st zone (the martensite zone, according to the XRD results (**Figure 8 b**)), is due to the martensite lattice "softening" according to Fedotov [**36, 37**] and Nakanishi [**39**].

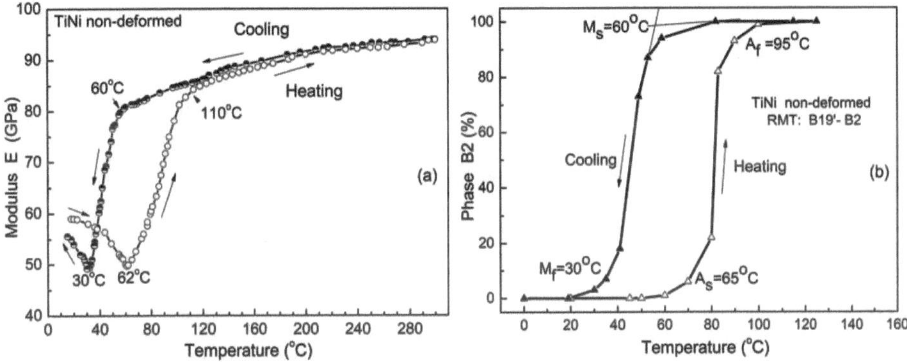

Figure 8. Comparison between critical temperatures of (a) E(T) graph and (b) RMT scheme (derived from DRX data in Figure 5) of the non-deformed Ti-54 wt. % Ni alloy, during heating and cooling.

In the 2nd zone (60-100 °C) B19′ martensite rearranges into B2-phase, so E and G rise abruptly. In the 3rd zone, at $100 < T < 700$ °C, the alloy, that had undergone RMT (B19′→B2), strengthened when the temperature increases. When the alloy is heated up to 700 °C, the moduli continue to grow revealing B2-phase stabilization. In the 4th zone, at T>700 °C, decrease of E and G arise from B2 crystal lattice stabilization till resistance to induced MT. For this alloy, the temperature of 700 °C is correlated with M_d, above which the B2 phase would not suffer RMT under stress.

3.3. Structure and properties of Ti-54 wt. %Ni after deformation

The samples of the same alloy composition were deformed at RT to obtain different prior plastic deformation, $\varepsilon_o = 2 - 30$ %. Tension curves of specimens with the different ε_o were practically identical (with their initial points coinciding), therefore some of them are shown in **Figure 9** as one "stress-strain" superposed plot. The dots on the curve denote the end of deformation and dashed lines – the off-loading of specimens. This load-extension diagram is typical for TiNi alloys with martensitic structure.

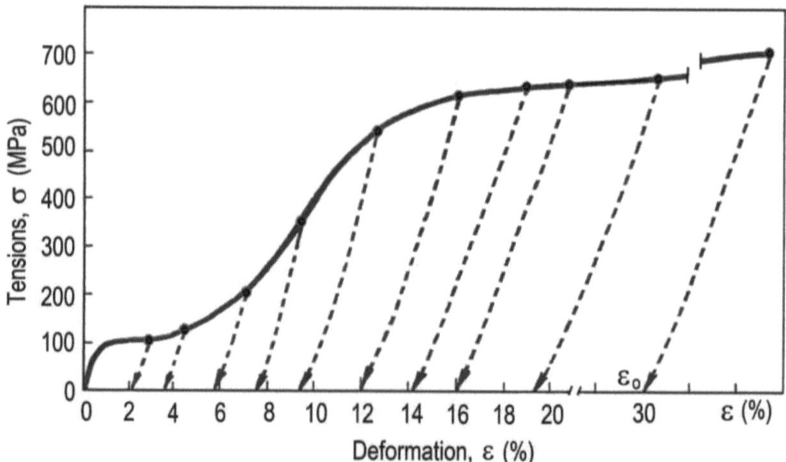

Figure 9. Ti-54 wt. %Ni alloy generalized tension curve.

XRD data show that in the structure of the alloy, deformed up to $\varepsilon_o = 2.2$ %, corresponding to the start of the pseudo-yield plateau, the martensite line $(020)_{B19'}$ moves towards smaller 2θ, and the others suffer some shifts **(Figure 10a, curves 1, 2)** [18, 19]. After prior plastic deformation to $\varepsilon_o = 5 - 7$ %, corresponding to the end of the plateau and initial stage of elastic deformation II and hardening, the structure changes substantially: additional new peak appears at $2\theta = 42 - 43^{\circ}$ that may be attributed to the $(110)_{B2}$ peak of high temperature B2 phase ($2\theta = 42.5^{\circ}$) or to the $(300)_R$ peak of the R phase ($2\theta = 42.55^{\circ}$) (**Figure 10a, curve 3**). When preliminary deformation increases up to 12 - 19 %, the new peak broadens and diminishes in intensity (**Figure 10, curves 4, 5**); with increasing of deformation up to $\varepsilon_o = 30$ % it appears again as the main peak, while martensitic peaks $(012)_{B19'}$, $(111)_{B19'}$, $(110)_{B19'}$ vanish, and the peak at $2\theta = 42 - 43^{\circ}$ broadens and splits, that is characteristic for the R phase [$(11\text{-}2)_R$ ($2\theta = 42.08^{\circ}$) and $(300)_R$ ($2\theta = 42.55^{\circ}$)] (**Figure 10a, curve 6**). The structure is unstable as if fluctuating and transforming from M to $R/B2_{def}$ $\rightarrow$to M_{def} $\rightarrow$to $R/B2_{def}$, thus accumulating deformation.

The diminution in intensity of the new line $(110)_{B2/R}$ during cooling of the deformed specimen with $\varepsilon_o = 6.9$ % was observed, as martensite lines $(020)_{B19'}$ and $(012)_{B19'}$ become more distinct (**Figure 10b**) [18, 19]. This indicates that the new peak belongs to the intermediate R phase coherent with both martensitic B19' and B2 phases. The structure restored during further heating to room temperature (see Figure 23c, 1st heating after 7%). The new peaks that appeared after deformation were identified as the peaks belonging to distorted high-temperature B2-phase, or R-phase, originally described by Dautovich and Purdy [45].

Considered as intermediate between B2-phase (CBC ordered of type CsCl [46-48]) and martensite B19' (monoclinic structure, space group P2$_1$/m [49-51]), the R-phase has a trigonal structure (space group P-3) [52-54], it is essentially a rhombohedral distortion of the cubic B2 austenitic phase, and its appearance is characterized by the increase in electric resistivity, splitting of the peak $(110)_{B2}$ in two near M_s, and anomalous changes in other physical parameters [4, 5, 7, 45, 54].

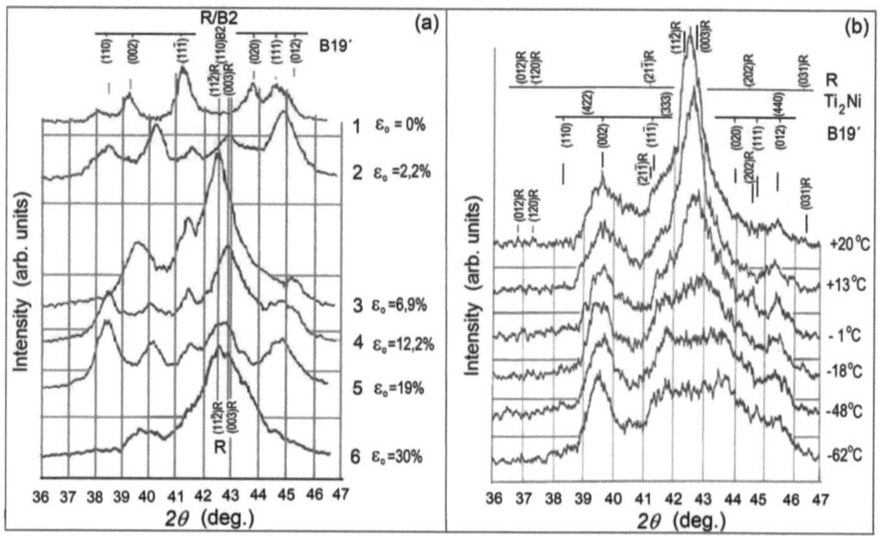

Figure 10. (a) Structural changes of the Ti-54 wt. % Ni alloy under deformation, (b) XRD of deformed TiNi alloy ($\varepsilon_0 = 6.9$ %) during cooling.

Within the pseudo-yield plateau and the initial stage of the subsequent elastic deformation II and hardening of TiNi (up to $\varepsilon_0 = 7$ %) deformation cause the proportional increase in resistivity ρ from $(82-84)\times10^{-8}$ $\Omega\cdot$m to $(98-100)\times10^{-8}$ $\Omega\cdot$m; but with the further growth of ε_0 to 19 % ρ changes very little (**Figure 11, curve 1**). The higher deformation, up to 30 %, causes the further increase of electrical resistivity ρ, up to $(107 \pm 2) \times 10^{-6}$ $\Omega\cdot$cm, due to plastic deformation and phase changes in the material.

After heating to 300 °C and cooling to RT, the resistivity ρ of TiNi specimens, deformed up to $\varepsilon_0 = 7$ %, is almost restored to the value of the undeformed alloy, while its structure returns to the martensite B19′, as a result of the removal of lattice distortion and realization of SME (**Figure 11, curve 2**).

The elastic modulus E, measured at RT (E_{20}), first falls ($\varepsilon_o = 2 - 3\%$) and then grows, even though the degree of previous deformation is increased to $\varepsilon_o = 5 - 7\%$ (**Figure 11, curve 3**), when R phase reveals itself in the structure. An increase in modulus is typical for the formation of intermediate phases (such as ω-phase, etc.) in titanium alloys [**36, 37**]. The values of E fall again as the plastic component builds up with $\varepsilon_o > 7\%$. In TiNi based alloys the high modulus R-phase is considered to be intermediate, while the unstable B2 phase is distinguished by the lower modulus E [**55**].

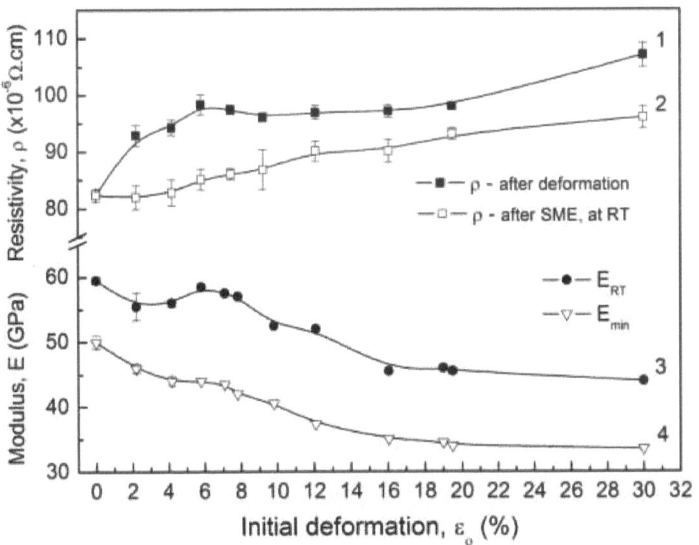

Figure 11. Physical properties of Ti-54 wt. % Ni alloy as the function of previous deformation, ε_o: (1) electrical resistivity (ρ) in deformed state and (2) ρ after heating up to 300 °C and realized SME; (3) variations of modulus E_{20C} and (4) E_{min}.

3.4. SME and RSME of Ti-54 wt. %Ni after deformation

DTA data for deformed TiNi alloy are shown in **Figure 12**. Endothermic peak shifts to the higher temperatures and becomes wider when the preliminary deformation increases, as indicated by Figures 6 (DTA of undeformed TiNi alloy) and 12, revealing additional phase transformations.

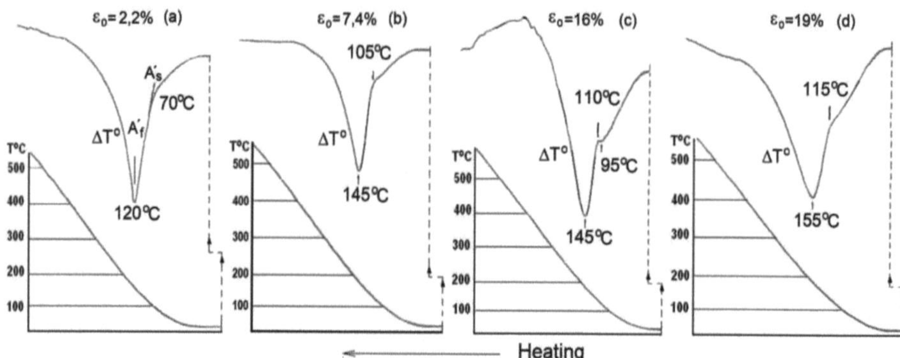

Figure 12. DTA results for the Ti-54 wt. % Ni alloy after tension to the different degrees of prior plastic deformation (ε_0): (a) 2.2 %, (b) 7.4 %, (c) 16 %, (d) 19 %.

Samples deformed by tension to different level of ε_0, were tested in the dilatometer, during 2-3 cycles of heating up to 300 °C and cooling (**Figure 13 a,b,c,d**). When comparing between dilatometric data for undeformed (see **Figure 6b**) and deformed TiNi alloy (**Figure 13**), it was established that the dilatometric curves changed radically after deformation, exhibiting SME and RSME of different magnitudes depending on ε_0.

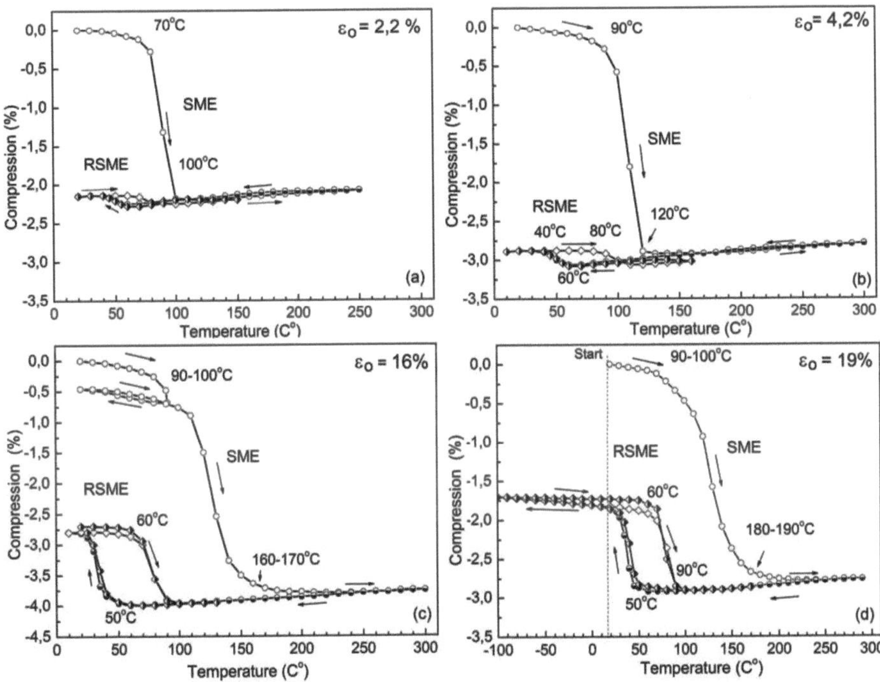

Figure 13. Some dilatometric curves of the Ti-54 wt. % Ni alloy deformed by tension to the different degree of prior plastic deformation (ε_o): (a) 2.2 %, (b) 4.2 %, (c) 16 %, (d) 19 % (V = 5 °/min)

Analyzing dilatometric data along with DRX, DTA results and E(T) for TiNi, deformed to $\varepsilon_o = 12$ % (**Figure 14 a,b,c**) we can pick out the stages: (**A-B**) and (**C-D**), whereon the extended specimen contracts slightly; (**B-C**), the stage of intensive shortening, in which the accumulated strain is recovered, revealing SME; and (**D -E**), the stage of thermal expansion of the specimen. Point **D** ordinate corresponds to the largest SME value ($\Delta h/h_o$) for the given strain. During cooling, after SME realization, initially the specimens become slightly shorter due to thermal contraction. When M_s is attained, the direct MT is accompanied by abnormal increase in the length of the specimen. The intensive endothermic reaction of RMT and contraction associated with the RSME is observed during repeated heating. **Figure 14** shows that deformation associated with SME shifts the endothermic peak to the higher temperatures, as well as the completion of the phase transformation (B19-R-B2). The modulus of elasticity behaves similarly (**Figure 14, d**). Along with this, the beginning of strain recovery (in unloaded state) and phase transformations occur in the alloy with low modulus of elasticity at the stage of softening. But the completion of transformation and shape restoration occurs with the significant increasing of modulus.

In the presence of stress, the strain recovery of alloy would be at even higher temperatures (when the elastic characteristics of the alloy are comparable with its response to the applied load). During second heating (or reheating), when RSME is realized, the reversible phase transformations and E(T) critical points are observed at lower temperatures shifted towards the critical temperatures of the initial alloy.

Under stress the strain recovery of alloy would be at even higher temperatures (when the elastic characteristics of the alloy are comparable with its response to the applied load). During second heating, when RSME is realized, the reversible phase transformations and E(T) critical points are observed at lower temperatures that shift toward the critical temperatures of the original undeformed alloy.

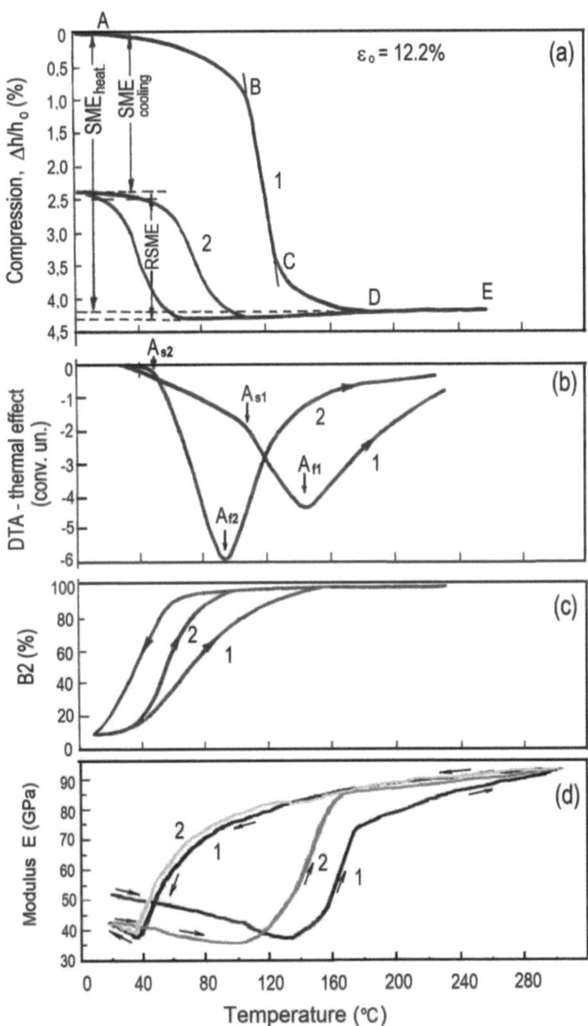

Figure 14. (a) dilatometric curves results, (b) DTA, (c) volume fraction of B2-phase (DRX), and (d) modulus E, for the Ti-54 wt. % Ni alloy, tested after prior plastic deformation of ε_0 12.2 % during two heating and cooling cycles, including (1) SME and (2) RSME.

When ε_0 rises to 7 % (the end of the pseudo-yield plateau), SME_{max} exhibit 4.6 - 4.9 % and falls afterwards (**Figure 15 curve 1**). Heating to room temperature lowers SME values significantly (**Figure 15 curve 2**), especially at $\varepsilon_0 > 7$ %. When ε_0 rises to 12 %, (the end of the stage of elastic deformation II and hardening, above the second yield point, in **Figure 1**), the RSME reaches its maximum (1.8 - 1.9 %), (**Figure 15 curve 3**). Similar results were observed in [**17**].

The largest value of recovery deformation corresponds to the highest recovery strain upon heating, as it was noted in [**4, 56**].

DTA and dilatometry data show that correlated with SME reversible deformation raises the starting and finishing points of the intensive endothermic reaction (A_s-A_f)(DTA); and the accumulated deformation recovers intensively (A_s-A_f) (SME) during the first heating (**Figure 16**). Up to $\varepsilon_0 = 7$ %, all three stages of recovery (A-D) are completed at the temperature corresponding to the endothermic reaction peak (A_f (DTA)). At $\varepsilon_0 > 7$ % the end-of-the-SME-temperature (A_f)(SME)) continues to rise, so that by A_f(DTA) only two first stages (A-B, B-C, in **Figure 14**) have time to take place, and the (C-D)-stage is included in the temperature range of the ascending branch of the DTA curve.

We notice that the temperatures of RSME realization are close to the RMT critical temperatures of undeformed alloy. Anyhow under severe strain SME (initial deformation being 20 - 30 %) critical temperatures and the transformation temperature decrease dramatically.

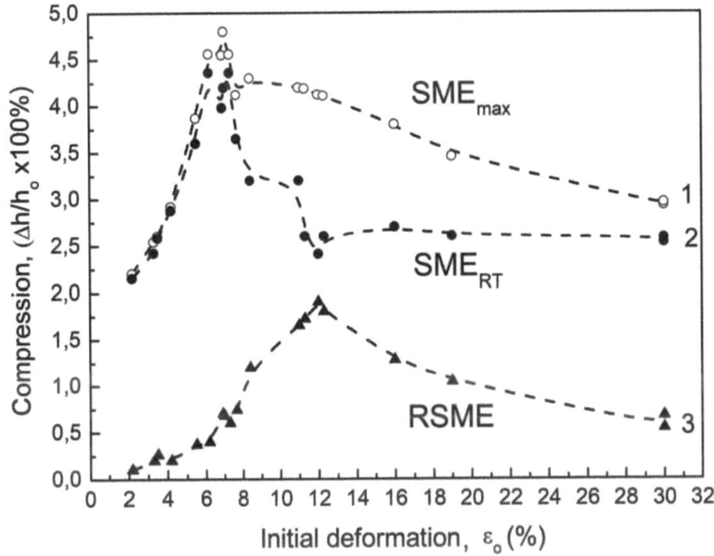

Figure 15. Prior plastic deformation dependence of: (1) SME_{max}, (2) SME_{20C}, and (3) RSME for Ti-54 wt. % Ni alloy.

This is because the effect of the true strain (irreversible, plastic deformation), $\square_p$, that accumulates intensively after the "elastic deformation-II and hardening" stage (**Figure 17a**). Its value was calculated as the difference between initial deformation, ε_o, accumulated by alloy after deformation and unloading (**Figure 9**), and the magnitude of SME_{RT} effect (at 20 ^{0}C) (Equation 7) or SME_{max} (Equation 8), where the magnitude of the RSME effect is already taken into account:

$$\varepsilon_p(1) = \varepsilon_o - SME_{20C} \qquad (7)$$

$$\varepsilon_p(2) = \varepsilon_o - SME_{max} \qquad (8)$$

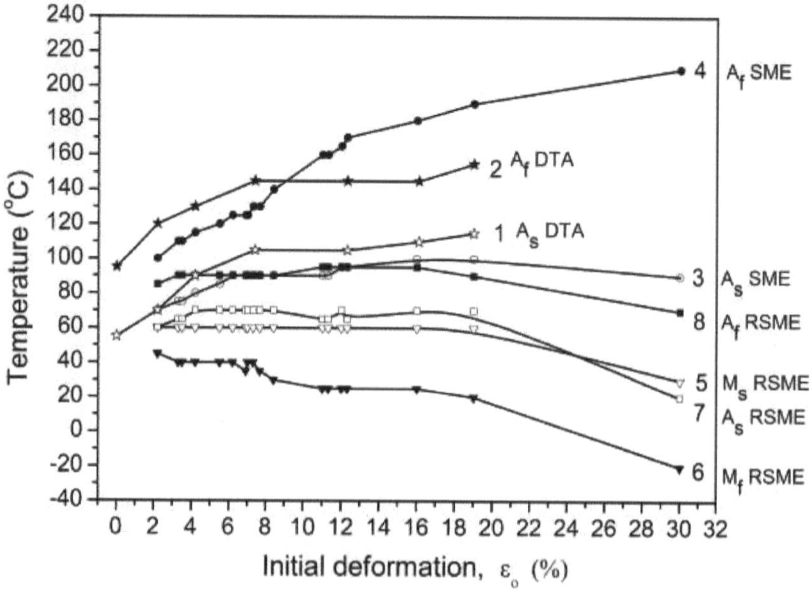

Figure 16. Effect of the prior plastic deformation (ε_o) for the Ti-54 wt. % Ni alloys on the critical temperatures of: (1, 2) DTA, (3, 4) SME and RSME, on cooling (5,6) and (7,8) heating.

Figure 17a shows that at $\varepsilon_o \approx 7$ % the plastic subcomponent of deformation reaches 2.2 %(ε_p2) or 2.8 %(ε_p1); and it rise sharply almost to 27.1 - 27.5 % at ε_o increasing up to 30 %. The TiNi density data speak in favor of the accumulation of structural defects (**Figure 17b**). The density of the deformed TiNi alloy changes depending on ε_o, more noticeably at $\varepsilon_o >12$ %, probably due to micro defects in its structure, resulting from the real plastic deformation; these structural defects adversely affect both effects, which strongly depend on the completeness of reversible phase transformations in the deformed SMA.

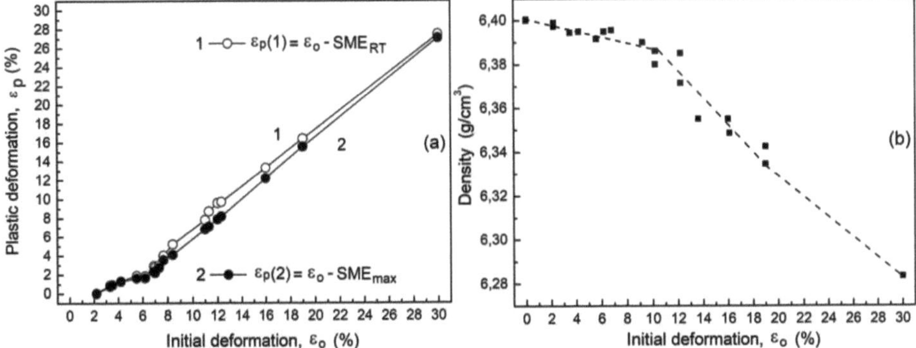

Figure 17. Prior plastic deformation dependence of: (a) irreversible plastic deformation (ε_p) accumulated in the alloy (not related to SME and RSME), and (b) the density of the deformed Ti-54 wt. % Ni alloy.

These results reveal severe pre-strain effect, namely ε_o, on the critical temperature intervals of SME/RSME. These proves that critical temperature data determined by DTA for strain-free alloy is not enough to set the heating mode for SME complete realization, even without loading.

3.5. Temperature changes of the elastic modulus E of Ti-54 wt. %Ni alloy after deformation

Plastic deformation does not alter the original E(T) pattern (**Figure 18 a, b**), but it induces T_{Emin} to shift towards the higher temperatures and to fall at the stage of "softening" (**Figure 11, curve 4**). Deformation up to 7 % shifts T_{Emin} noticeably (**Figure 18 a**), but at the higher deformation it changes slightly (**Figure 18 b**). The greater deformation the lower modulus, provided unrelated to SME plastic irreversible deformation accumulates in alloy; temperature range of modulus ascending branch widens, up to 19 % T_{Emin} varies insignificantly, decreasing in the most deformed alloy (30 %).

Figures 19 a, b shows that E_{min} decreases when the initial deformation ε_0 grows while cooling or during the 2nd heating of specimens.

The largest moduli displacements to elevated temperatures are observed during the primary heating of TiNi deformed alloys, as shown in **Figure 20**.

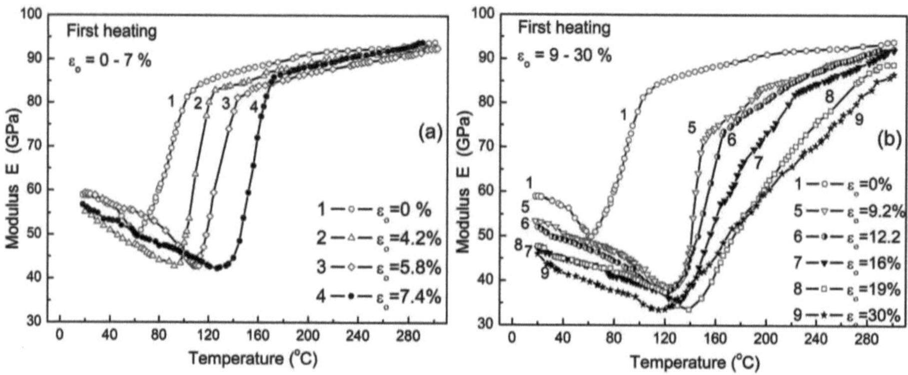

Figure 18. Temperature changes of the elastic modulus E of Ti-54 wt. %Ni alloy in its initial state (ε_0=0) and deformed state: (a) ε_0=4.2 %, 5.8 %, 7.4 %, (b) (ε_0 = 9.2 %, 12.2 %, 16 %, 19 % and 30 %), during primary heating (when SME is realized).

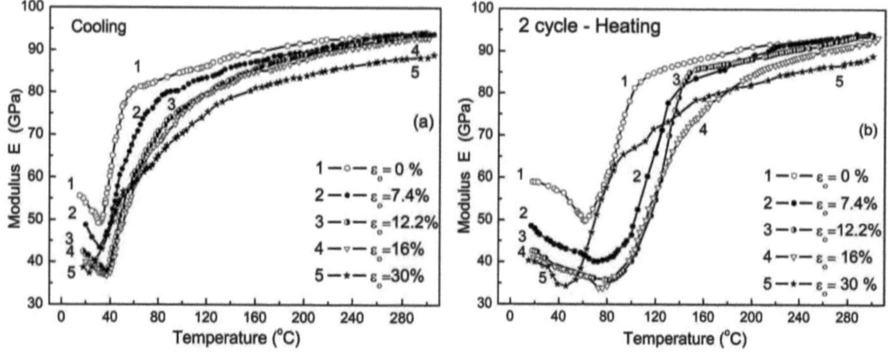

Figure 19. Comparison between the modulus E changes the Ti-54 wt. % Ni alloy, in initial and deformed state, (a) at the cooling 300 °C, and (b) 2nd heating up to 300 °C (when RSME is realized).

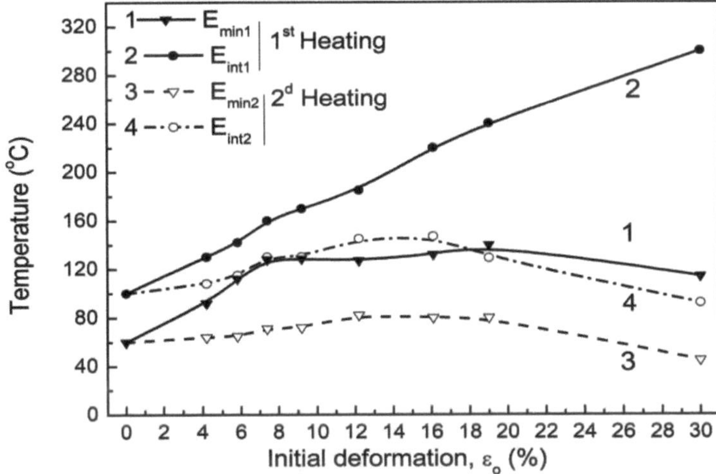

Figure 20. Changes in the T_{Emin} and T_{Eint} position for the deformed alloy Ti-54 wt. % Ni, on the first and second heating up to 300 °C.

4. Discussion

When comparing between E(T) curves on the 1st heating (when TiNi-alloy undergoes SME, see Figure 18) and upon cooling and reheating (when RSME is realized, see Figure 19), we note that the course of changing of the values of E on the 2nd heating follows the same route as during the 1st heating, but the temperature shift of E_{min} is less significant. After the 2nd heating, moduli behavior under cooling was identical to that after the 1st heating, so they are not given here. On the 1st heating the alloy undergoes SME-associated transformations; then upon cooling and re-heating, the T_{Emin} decreases to the initial value of undeformed alloy.

Comparing together DTA, E (T), and SME data on the critical temperatures for TiNi alloy in initial state and after deformation to different ε_o (**Figure 21**), we can pick out 4 characteristic strain intervals. In the first interval, up to 4 %, the SME (**Figure 21,**

curves 1 and 2) occurs within endothermic DTA effect (**Figure 21, curves 3 and 4**), and in the range of sharp increase in the elastic modulus from E_{min} to E_{int} (**Figure 21, curves 5 and 6**); T_{Eint} coincides with the completion of the DTA effect. In the second interval, $\varepsilon_o = 4 - 7\ \%$, T_{Emin} increases and approaches the SME completion temperature, i.e. the deformation is partly restored, given lattice softening; it is peculiar to reversible phase transformation initiated by preliminary deformation. In the third interval, $\varepsilon_o = 7 - 12\ \%$, the recovery of the accumulated SME deformation also begins in the softened alloy; the completion of the SME is observed at the temperatures, that exceed the completion of the DTA reaction, within the temperature interval of intensive increase in the modulus of elasticity. At the large preliminary deformations, in the fourth interval, the gap between T_{Emin} and T_{Eint} increases dramatically, i.e. SME realization is stretched in temperature and SME value falls sharply (see **Figure 15**).

In the second cycle (**Figure 22**) temperatures T_{Emin2} and T_{Eint2} can be compared with the critical temperatures of the RSME. We noted that the temperatures both $(A'_s\text{-}A'_f)$/RSME on heating and M'_s/RSME on cooling, are less sensitive to prior deformation (up to $\varepsilon_o = 19\ \%$) contrary to M'_f/RSME, that decreases markedly starting with $\varepsilon_o = 7\text{-}8\ \%$. Upon reheating of the deformed alloy (with ε_o up to $12\ \%$) the temperatures T_{Emin} and T_{Eint} grow; so that starting from $\varepsilon_o = 8\ \%$ the $(A'_s/\text{RSME} < T_{Emin2} < A'_f/\text{RSME} < T_{Eint2})$ interrelationship is observed; that is, the change in the RSME shape upon heating starts at reduced elasticity modulus (**Figure 22, curves 1 and 3**). Nevertheless, the RSME completes with a sharp increase of the elastic properties of the alloy (**Figure 22, curves 2 and 4**). At $\varepsilon_o > 19\ \%$ all the aforementioned critical temperatures are markedly reduced.

Comparing together dilatometry data, temperature variation of the elastic modulus E, and structural changes of the alloy, deformed to $\varepsilon_o = 7\ \%$ and $30\ \%$ (**Figure 23 and 24**), it was revealed that notable T_{Emin} shift results from accumulation of structural changes correlating with the reversible SME deformation; while the absolute value of the E depends on the residual plastic deformation due to internal oriented micro-stresses retained by the alloy.

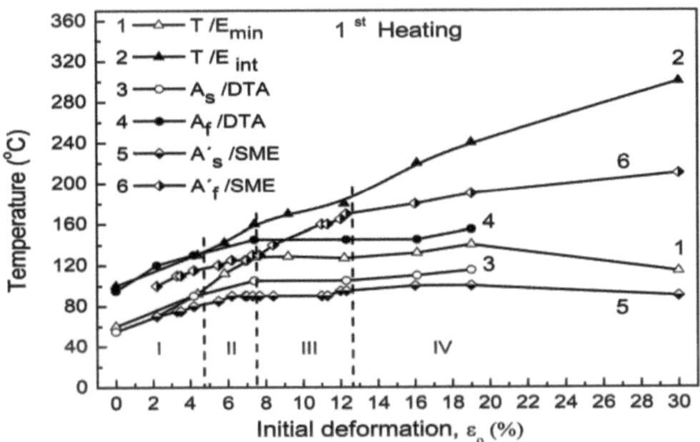

Figure 21. Critical temperatures of the modulus (E_{min} and E_{int}), DTA (A_s and A_f), and SME (A$'_s$ and A$'_f$) for the Ti- 54 wt. % Ni alloy, deformed up to ε_0, determined during 1st heating up to 300 ^{o}C (when SME is realized).

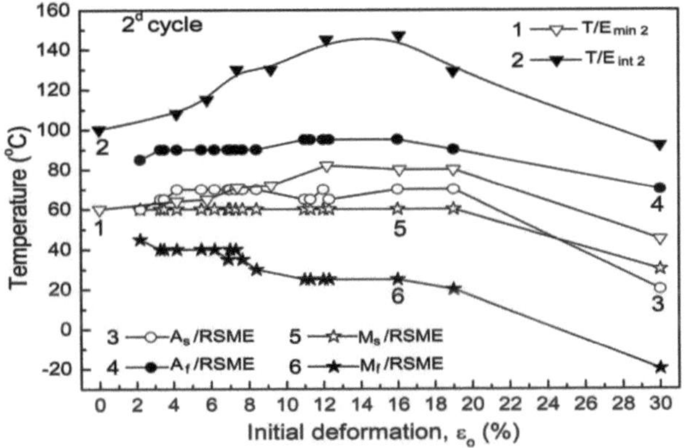

Figure 22. Critical temperatures of the modulus (E_{min2} and E_{int2}) and RSME (M$'_s$ - M$'_f$ / A$'_s$ - A$'_f$) for the Ti-54 wt. % Ni alloy, initially deformed up to ε_0, determined during 2nd heating up to 300 ^{o}C and cooling (when SME is realized).

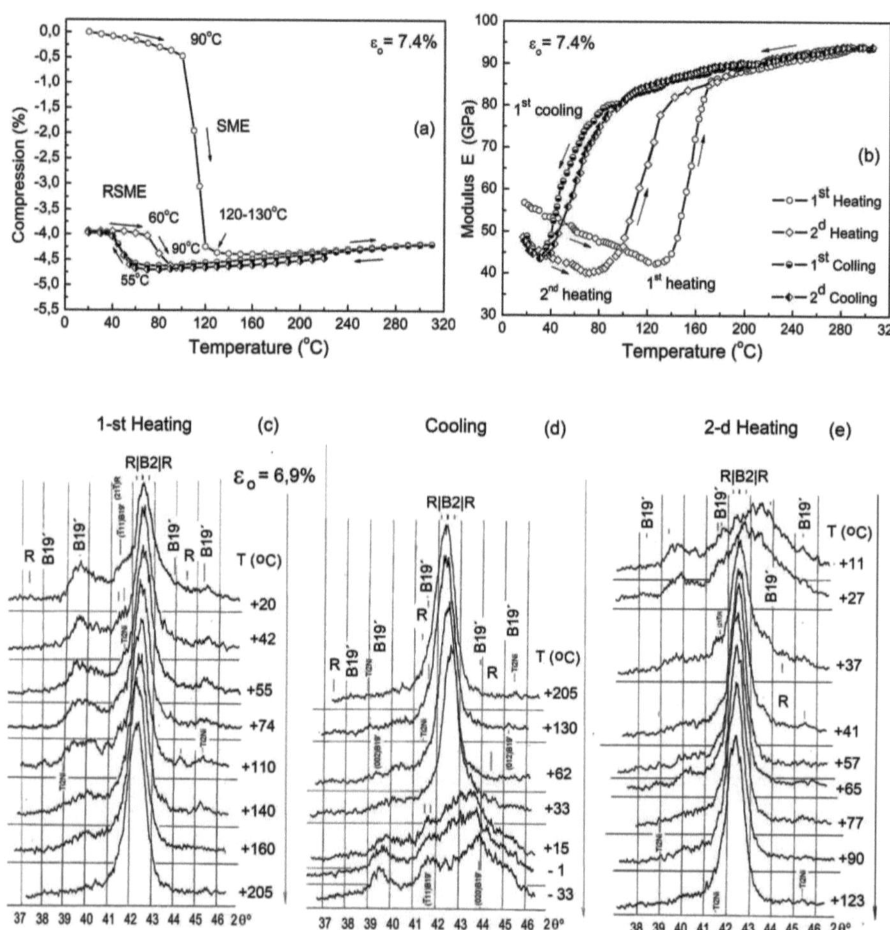

Figure 23. (a) dilatometry data, (b) elastic modulus temperature variation, and XRD data for: (c) the first heating , (d) cooling and (e) second heating of Ti-54 wt. % Ni alloy, pre-deformed up to ε_o = 6.9 and 7.4 %.

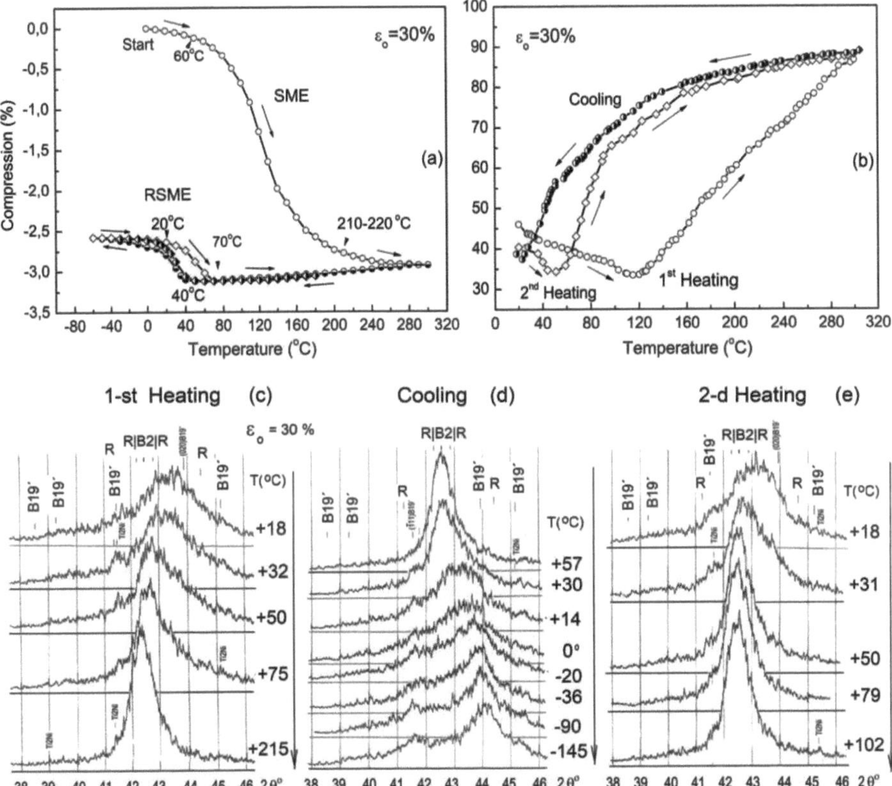

Figure 24. (a) the dilatometry data, (b) elastic modulus temperature variations, and the XRD data of structural change for: (c) the first and (d) second heats of the Ti-54 wt. %Ni alloy, pre-deformed to $\varepsilon_0 = 30$ %.

Moreover, in pre-strained alloy (up to $\varepsilon_0 = 30$ %) T_{Emin} decreases significantly upon second heating, since the deformation stabilize the high-temperature R/B2 phase (**Figure 24**). XRD thermal analysis of the specimen during the 1st and the 2nd heating shows that strong structural distortions, accumulated during pre-deformation, are basically remain unchanged and don't allow to restore the initial structure of alloy after 2 cycles of heating (to 300 ^{o}C) and cooling. Within critical temperature interval this alloy notably reveals the R-distortion, during RMT: B2$\leftrightarrow$R$\leftrightarrow$B19′ (**Figure 24**). The obtained data can be successfully explained using the theoretical model for the atom displacements that occur during the B2$\rightarrow$R and R$\rightarrow$B19 transformations proposed by Goryczka, H. Morawiec [**53**]. According to this model "a coordinated atom shuffle appears in the {110}B2 plane and the <110>B2 direction. The value of the atom shift equals 3 % with respect to the parent phase cell parameter. Shearing of the trigonal cell in the {111}R plane and the 112 R direction is the final transformation step to "monoclinic martensite". This fact explains the observed smooth transformation of one phase into another in a severely deformed alloy, where oriented in the stress field phases are completely coherent. The accumulated deformation stabilizes the transitional R-phase thus extending its temperature field and shifting the SME interval (R$\rightarrow$B2) to the higher temperatures on heating, and RSME interval (R$\rightarrow$B2) to the lower temperatures on cooling.

5. Conclusions

The structural and dynamic elastic modulus (E) changes, as a function of temperature and deformation up to 30 %, as well as the parameters of SME and RSME of initially B19 martensite TiNi alloy, were investigated. The following conclusions are drawn:

(1) The investigated Ti-54 wt. % Ni alloy exhibits low values of moduli (E and G) that become negligibly small at the critical points of RMT. Accumulation of the reversible deformation in TiNi (recoverable upon heating in the course of SME) occurs owing to low lattice stability of the initial martensite phase B19′.

(2) During deformation within the "pseudo-yield" plateau, the initial B19′ phase is distorted to its limit until the formation of preferentially oriented new coherent phase, thus the SME-deformation in the alloy; the recoverable SME-deformation is restored by lattice distortion and transformation of B19′ to a high-temperature R phase, with preferential lattice orientations.

(3) Increase of resistivity ρ and increase of the SME reversible deformation value, in proportion to pre-deforming until the ultimate extent ($\varepsilon_o = 7 - 8$ %) are associated with the RMT under stress (B19$\rightarrow$R), leading to the formation of specific distorted structure and the coherent R phase, developed at the stage of the "pseudo-yield" plateau. This distortion disappears after the first heating of deformed TiNi alloy to the temperature exceeding A_f.

(4) SME deformation, retaining in loaded-unloaded alloys due to its accommodation, thus require additional energy for its recovery. This leads to growing of the temperatures, at which RMTs occur and the shape begins to get restored. It also promotes the increase of resistivity ρ and the value of the SME reversible

deformation, <u>in proportion</u> to pre-deforming until the ultimate extent ($\varepsilon_0 = 7 - 8\,\%$) is reached.

(5) While the distortions giving rise to the SME are caused by the applied stress, similar RSME lattice distortions are due to internal oriented stresses, fixed as a result of elastic-plastic deformation at the strengthening stage (up to $\varepsilon_0 = 12 - 13\,\%$, for TiNi); on cooling, these internal stresses preferentially orient the martensitic shifts when the elastic moduli diminish drastically during RMT; thus some macro-deformation of the RSME accumulates, on cooling, and restore, on heating, while E is growing. Constancy of these internal stresses determines the stability of the RSME parameters at thermo-cycling.

(6) Elastic and plastic deformations and structure changes, under applied stress, in polycrystalline TiNi alloy, are realized and accumulated to various degrees in individual grains because of their different orientations relative to the applied force; these make them not equally stable in respect to phase transformations during heating and cooling, by expanding the ranges of realization of SME and RMT; this has especial effect on the shift of A_f (SME) and M_f (RSME) in TiNi alloy deformed with ε_0 more than 12 %.

References

1. Chang, L.C.; Read, T.A. Plastic deformation and diffusionless phase change in metals – The Gold-Cadmium beta phase. *J. Transactions of AIME* **1951**, *191*, 47. [Google Scholar]

2. Buehler, W.J; Gilfrich, L.W.; Wiley, R.C. Effect of low-temperature phase changes on the mechanical properties of alloys near composition TiNi. *J. of Applied Physics* **1963**, *34* N5, 1475-1477. [Google Scholar]

3. Kurdyumov, G.V.; Khandros, L.G. On the "thermoelastic" equilibrium on martensitic transformations. In *Dokl. Akad. Nauk SSSR.* **1949**, *66*, No. 2, 211-214. [Google Scholar]

4. Perkins, J. Thermomechanical Characteristics of Alloys Exhibiting Martensitic Thermoelasticity. In Shape memory effects in alloys. (Ed. by Jeff Perkins). The Metallurgical Society of AIME. Proceedings of the International Symposium on Shape Memory Effects and Applications. Toronto, 1975. Plenum Press. New York and London. 583 p. ISBN 978-1-4684-2211-5. [CrossReff]

5. Otsuka, K; Wayman, C.M; Eds; Shape memory Materials. 1999; Cambridge University Press. 284 p. ISBN 0-521-44487-X. [Google Scholar]

6. Wasilewski, R.J. On the "reversible shape memory effect" in martensitic transformation. *Scripta Metalurgica.* **1975**, *9*, 417-422 [Google Scholar]

7. Otsuka, K.; Ren, X. Physical metallurgy of Ti-Ni- based shape memory alloys. *Progress in Materials Science.* **2005**, *50*, 511–678 [Google Scholar]

8. Miyazaki, S.; Kim, H.Y.; Hosoda, H. Development and characterization of Ni-free Ti-base shape memory and superelastic alloys. *Materials Science and Engineering: A.* **2006**, *438*, 18-24. [Google Scholar]

9. Perkins, J. Martensitic substructural prerequisites for shape memory effect (SME) behavior. *Scripta Met.* **1975**, 9, 121-127. [Google Scholar]

10. Belousov, O.K. Temperature dependence of physical properties and connection of transformation in TiNi with phase diagram. *Izvestiya Akademii Nauk SSSR,* **1981,** *Metally,* 240-242. [Google Scholar]

11. Wang, F.E.; Buehler, W.J., Pickart, S.J. Crystal Structure and a Unique "Martensitic" Transition of TiNi. *Journal of Applied Physics*, **1965**, *36*, 3232-3239. [Google Scholar]

12. Gefen, Y.; Rozen, M. Behavior of the elastic and inelastic properties of Au-47.5at.%Cd in the vicinity of the cubic - orthorhombic phase transformation. *Philosophical magazine.* **1972**, 26, 727-736. [CrossRef]

13. Melton, K.N.; Mercier, O. Deformation behavior of NiTi-based alloys. *Metallurgical transactions A.* **1978**, *9*, 1487-1488. [CrossRef]

14. Nagasawa, A. A new concept on the shape memory effect in metals and alloys. *Physica Status Solidi (a).* **1971**, 8, 531-538. [Google Scholar]

15. Kornilov, I.I.; Belousov, O.K.; Kachur E.V. Titanium Nickelide and Other Shape Memory Alloys. Publ Nauka, Moscow. 1977.[in Russian] [Google Schoolar]

16. Kovneristy, Yu.K.; Fedotov, S.G.; Matlakhova, L.A. Effect of deformation on phase transformations and modulus of elongation in titanium nickelide – based alloy. In Titanium Science and Tecnology. Proc. of the 5th International Conference on Titanium. Congress-Center, Munich, FRG, September 10-14, 1984. Volume 3.; Lütjering, G; Zwicker, U; Bunk. W. Deutsche Gesellschaft für Metallkunde EV. v.3. p.1673-1681. ISBN 3883550833, 9783883550831 [Google Books]

17. Kovneristy, Y.K.; Fedotov, S.G.; Matlakhova, L.A.; Oleynikova S.V. Shape memory and shape reversibility effects in a TiNi alloy as a function of deformation. *Fiz. Met. Metallogr.(USSR).* **1986,** *62*, 124-129. [Google Scholar]

18. Kovneristy Yu.K., Fedotov S.G., Matlakhova L.A. The influence of plastic deformation on the structure, shape memory effect and other properties of TiNi alloy. In: Shape Memory Alloy 86 - Proceeding of the International Symposium CN Shape Memory Alloys. (Ed. by T.Y. Hsu and T.Ko Chu Youyi) sept. 6-9, 1986. Guilin, China. China Academic Publishers (1986) p. 175-180.

19. Fedotov, S.G.; Kovneristy, Yu.K.; Matlakhova, L.A.; Zhebynieva N.F. Structural changes in alloy TiNi with the shape memory effect during deformation. *Physics of Metals and Metallography (Fiz. Metal. Metalloved)* **1988**, *65*, 564-569. [Google Scholar]

20. Saburi, T.; Takagaki, T.; Nenno, S.; Koshino, K. Mechanical Behavior of Shape Memory Ti-Ni-Cu Alloys. In *Proceedings of the MRS International Meeting on Advanced Materials.* **1988**, 9, 147-152. [Google Scholar]

21. Otsuka, K.; Sakamoto, H.; Shimizu, K. Successive stress-induced martensitic transformations and associated transformation pseudoelasticity in Cu-Al-Ni alloys. *Acta Metallurgica,* **1979**, *27*, 585-601. [Google Scholar]

22. Saburi, T.; Yoshida, M.; Nenno, S. Deformation behavior of shape memory TiNi alloy crystals. *Scripta Metallurgica,* **1984**, *18*, 363–366. [Google Scholar]

23. Perkins, J; Edwards, G.R.; Such, C.R.; Johnson, J.M., Allen R.R. Thermomechanical Characteristics of alloys exhibiting martensitic thermoelasticity. p. 273-303. In: Shape memory effects in alloys. (Ed. by Jeff Perkins). The Metallurgical Society of AIME. Proceedings of the International Symposium on Shape Memory Effects and Applications. Toronto, **1975**. Plenum Press. New York and London. 583 p. ISBN 978-1-4684-2211-5. [CrossRef]

24. Krishnan, R.V; Brown, L.C. Pseudoelasticity and the Strain-Memory Effect in an Ag-45 at.% Cd Alloy. *Metallurgical and Materials Transactions B.* **1973**, *4*, 423-429 [Google Scholar]

25. Jackson, C.M; H.J. Wagner, H.J; Wasilewski R.J. 55-Nitinol -The Alloy with a memory: Its Physical metallurgy, properties and applications. *NASA Report SP-5110.* **1972**. 94 p. [CrossRef]

26. Wasilewski, R.J. The effects of applied stress on the martensitic transformation in TiNi. *Metallurgical and Materials Transactions.* **1971**, *2*, 2973-2981. [Google Scholar]

27. Silva, R.J.; Matlakhova, L.A.; Matlakhov, A.N.; Pereira, E.C.; Monteiro, S.N.; Rodriguez, R.S. Thermal Cycling Treatment and Structural Changes in Cu-Al-Ni

Monocrystalline Alloys. In *Materials Science Forum*. **2006**, *514*, 692-696. Trans Tech Publications. [Google Scholar]

28. L.A. Matlakhova, E.C. Pereira, A.N. Matlakhov, S.N. Monteiro. Chapter 4: Structure and properties of a monocrystalline Cu-Al-Ni alloy submitted to thermal cycling under load, In: Shape Memory Alloys: Manufacture, Properties and Applications. Ed. Chen, H.R. (2010) p. 113-143. Editora Nova Science Publishers, USA [Google Scholar]

29. Wayman, C.M.; Shimizu, K. The Shape Memory ('Marmem') Effect in Alloys. *Metal Science Journal*. **1972**, *6*, *175*-183. [Google Scholar]

30. Borisova, S.D.; Monasevich, L.A.; Paskal' Y.L. Crystallographic Calculation of the Reversible Deformation in the Shape Memory Effects of Titanium Nickelide. *Metallofizika*, **1983**. *5*, 66–70. [Google Scholar], [SPRINGER]

31. Wasilewski, R.J. On the nature of the martensitic transformation," *Metallurgical Transactions A*, **1975**, *6*, 1405–1418. [Google Scholar]

32. Kurdyumov, G.V.; Martensitic transformations (Review). *Metallofizika*. 1979. *1*, 81-91. [CrossRef]

33. Matlakhova, L.A., Motta, A.C., Matlakhov, A.N., Kolmakov, A.G., Sevostianov M.A. Módulo de elasticidade da liga monocristalina Cu-Al-Ni deformada por compressão. In proceedings of: 18º Congresso Brasileiro de Engenharia e Ciência dos Materiais - CBECiMat; 24-28/11/2008, Porto de Galinhas – PE, Brasil, CD, p. 6463-6474.

34. Callister, W.D.; Rethwisch, J.D.G. *Material science* and *engineering. An Introduction*. 9th Edition. Eds. Callister, W.D.; Rethwisch, J.D.G. **2013**. John Wiley&Sons, New York, 936 p. ISBN: 978-0-471-73696-7. [CrossRef]

35. Gefen, Y.; Rozen, M. Behavior of the elastic and inelastic properties of Au-47.5at.%Cd in the vicinity of the cubic - orthorhombic phase transformation. *Philosophical magazine*. **1972**, *26*, 727-736. [CrossRef]

36. Fedotov, S. (1973). Peculiarities of changes in elastic properties of titanium martensite, in R. Jaffee & H. Burte (eds), Titanium Science and Technology (Proc. 2nd Int. Conf. on Titanium), Plenum Press, New York, pp. 871–881. [CrossRef]

37. Fedotov S.G. Dependence of the elastic properties of titanium alloys on their compositim and structure. In Titanium and its Alloys. I.I.Kornilov, Ed. Akademiya Nauk SSSR, (1964) (transl. Israel Program for Scientific Translations Ltd., IPST. Cat. No. 1454, 1966, p. 199-215]. [Google Scholar]

38. Lobodyuk, V.; Estrin, E.I. Martensitic transformations. Cambridge International Science Publishing Ltd (2014) 375 p. ISBN-13: 978-1-907343-99-5, [CrossRef]

39. Nakanishi. N. Lattice softening and the origin of SME. pp. 147-1776. In: Shape memory effects in alloys. Ed. J. Perkins. Metallurgical Society of AIME. Plenum Press. New York and London (1975) [Google Scholar]

40. Mercier, O.; Melton, K.N.; Gremaud, G.; Hagi, J. Single-crystal elastic constants of the equiatomic NiTi alloy near the martensitic transformation. *Journal of Applied Physics*. **1980**, *51*, 1833-1834. [Google Scholar]

41. Moine, P.; Michal, G.M.; Sinclair R. A morphological study of "Premartenstic" effects in TiNi. *Acta metallurgica*. **1982**, *30*, 109-123. [CrossRef]

42. Michal, G.M.; Moine, P.; Sinclair R. Characterization of the lattice displacement waves in premartensitic TiNi. *Acta metallurgica*. **1982**, *30*, 125-138. [Google Scholar]

43. Clapp, P.C. A localized soft mode theory for martensitic transformations. *Physica status solidi (b)*. **1978**, *57*, 561-569. [Google Scholar]

44. Fedotov, S.G. The Shape Memory Effect and the Superelasticity of Alloys. In *Dokl. Akad. Nauk SSSR*. **1986**, *290*, 1115-1118. [Google Scholar]

45. Dautovich, D.P.; Purdy, G.R. Phase transformation in TiNi. *Canadian Metallurgical Quarterly* **1965**, *4*, 129-143. [Google Scholar]

46. Margolin, H.M.; Ence, E.; Nielsen J.P. Titanium-nickel phase diagram. *Trans. AIME*. **1953**, *197*, p.243-247. [CrossRef]

47. 43. Poople, D.M.; Hume-Rothery W. The equilibrium diagram of the system nickel-titanium. *J. Inst. Metals*. **1954**, *83*, 473-475. [Google Scholar]

48. Purdy, G.R. Parr, J.G., A study of the titanium-nickel system between Ti_2Ni and TiNi. Trans. Met. Soc. AIME 1961, 221, 636 [Google Scholar]

49. Kudoh, Y; Tokonami, M.; Miyazaki S.; Otsuka K. Crystal structure of the martensite in Ti-49.2 at.%Ni alloy analyzed by the single crystal X-ray diffraction method. *Acta Metall*. **1985**, *33*, 2049-2056. [CrossRef]

50. Buhrer, W.; Gotthardt, R.; Kulik, A.; Mercier, O.; Staub, F. Powder neutron diffraction study of nickel-titanium martensite. *J. Phys. F*. **1983,** *13,* L77-L82 [CrossRef]

51. Huang, X.; Ackland, G.J.; Rabe, K.M. Crystal structures and shape-memory behaviour of NiTi. *Nature materials*. **2003**, *2*, 307-311. [Google Scholar]

52. Fukuda, T.; Saburi, T.; Doi, K.; Nenno, S. Nucleation and self-accommodation of the R-phase in Ti-Ni Alloys. *Materials Transactions, JIM*, **1992**, *33*, 271-277. [Google Scholar]

53. Goryczka, T.; Morawiec, H. Structure studies of the R-phase using X-ray diffraction methods. *Journal of Alloys and Compounds*, **2004**, 367, 137-141.[CrossRef]

54. 47. Miyazaki, S. Martensitic transformation in TiNi alloys. *In*: Miyazaki, S., Fu. Y. Q., Huang, W. M. *Thin Film Shape Memory Alloys Fundamentals and Device Applications*. Cambridge University Press. **2009**, 73-87. [Google Scholar], [Google Books]

55. Fedotov, S.G.; Matlakhova, L.A.; Zebyneva, N.F.; Kovneristyi, Y.K. The effect of TiNi-alloy composition on SME, RSME, and other physico-mechanical properties. *Technology of light alloys (in Russian)*. **1990**, *4*, 24-28.

56. Kovneristyi, Y.K.; Matlakhova, L.A.; Matveeva, N.M.; Kostyanaya, O.V. Characteristics of the Shape Memory Effect in Rapidly Quenched TiNi-TiCu Alloys. *Izv. Akad. Nauk SSSR, Met.*. **1988**, 138-142. [WOS]

An accurate and non-destructive solution for elastic moduli and damping measurement

Pereira, A. H. P.
ATCP Engenharia Física

Sonelastic® is a set of configurable systems for the non-destructive characterization of damping and elastic moduli (Young's modulus, shear modulus, and Poisson's ratio) employing the Impulse Excitation Technique, which is based on natural frequencies of vibration. When submitted to a light mechanic impact, the sample under test emits a characteristic sound according to its dimensions, mass and elastic properties. The frequencies and decay rate of the acoustic response allow an accurate determination of the elastic moduli and damping. Sonelastic® measures the elastic moduli of any rigid material in the shape of discs, rings, rectangular or cylindrical bars with dimensions ranging from 20 millimeters (3/4 inch) to 5.3 meters (17.4 feet). The typical configuration comprises a software, an acoustic sensor and a sample holder that will vary according to the sample's geometry and dimensions. Accessories, such as the automatic electromagnetic impulse device and instrumented furnaces, allow automatic measurements as a function of time and temperature. The Sonelatic® systems are in accordance with E1876, E756,
C1548 and C1259 ASTM standards.

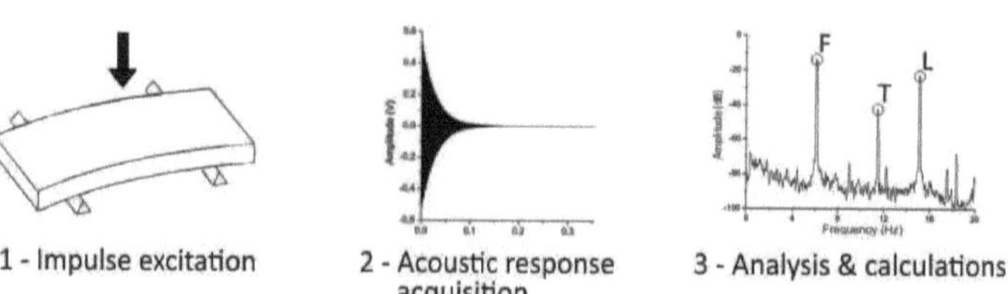

1 - Impulse excitation   2 - Acoustic response acquisition   3 - Analysis & calculations

Impulse Excitation Technique basic steps. More information at www.sonelastic.com.

The accurate and non-destructive solution for elastic moduli and damping measurement.

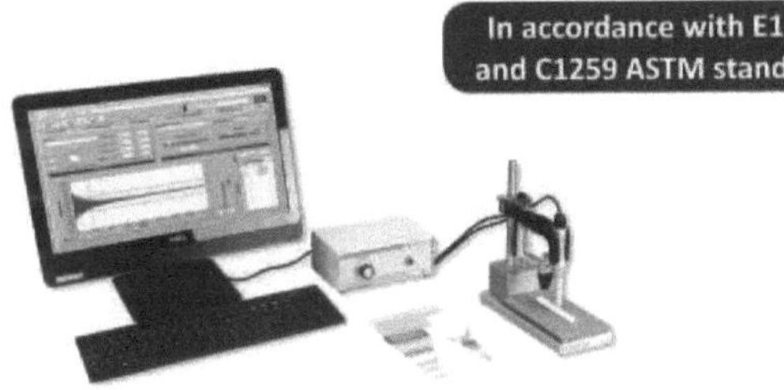

Room temperature

High temperatures

The Impulse Excitation Technique:

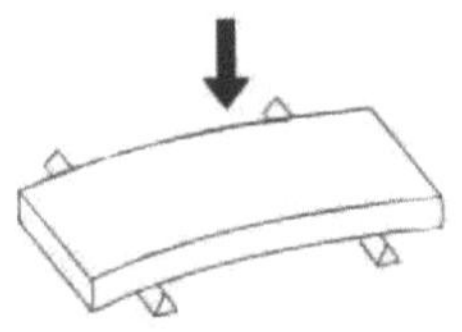

1 - Impulse excitation

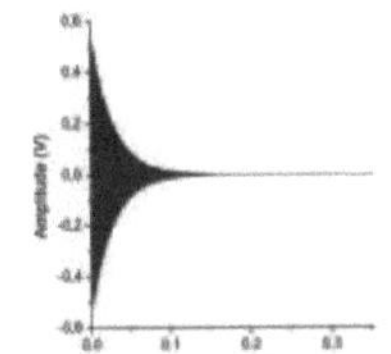

2 - Acoustic response acquisition

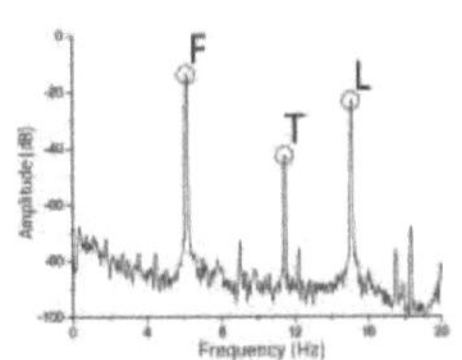

3 - Analysis & calculations

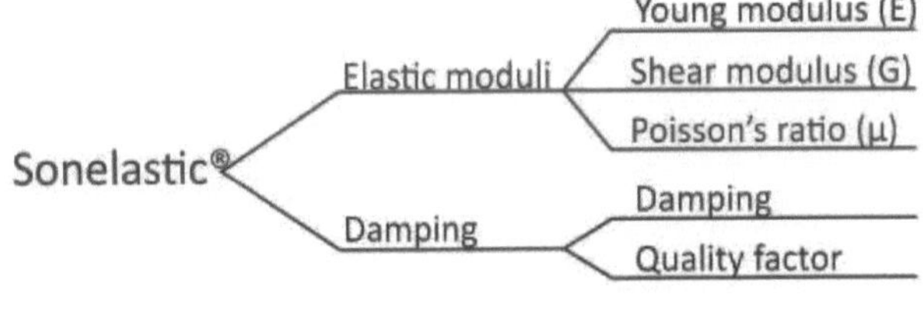

www.sonelastic.com

Printed by Books on Demand GmbH, Norderstedt / Germany